D1212954

AIR

Gabrielle Woolfitt

MAC KILLOP SCHOOL LIBRARY

Wayland

Titles in this series
Air
Earth
Fire
Water

TOPIC CHART

	SCIENCE*	ENGLISH	MATHS*	TECHNOLOGY	GEOGRAPHY	HISTORY	PHYSICAL EDUCATION	ART	RELIGIOUS EDUCATION
What is air?	AT 1 L2–5 AT 3 L4–5	AT 1 L2–5 AT 2 L3							
Gases in the air	AT 2 L2–5 AT 3 L4–5	AT 2 L4–5 AT 3 L3–4	ATS 1,2,4,5 L2–5						
Breathing	AT 2 L3–5								
Smell	AT 3 L4	AT 3 L2–5							
Spirits and angels		AT 1 L2–5 AT 3 L2–5							✔
Winds and hurricanes					AT 2 L4	AT 3 L2–5			
Air power	✔			AT 1–4, L2–5		CSU 3,4			
Birds and bees	AT 2 L2–5								
Air travel	AT 3 L4–5 AT 4 L2–5			AT 4 L4–5	AT 4 L3	CSU 3,4			
Air sports				AT 4 L4–5			✔		
Air pollution	AT 3 L2–5				AT 5 L2–5				
Kites and streamers				AT 1–4, L2–5			✔	✔	
Making a kite				AT 1–4, L2–5					
Balloon messages	AT 3 L4	✔			AT 2 L2–5				

KEY CSU = Core Study Unit AT = Attainment Target L = Level * Proposed ATs, October 1991

First published in 1992 by
Wayland (Publishers) Ltd
61 Western Road, Hove
East Sussex BN3 1JD, England

© Copyright 1992 Wayland (Publishers) Ltd

Editor: Cath Senker
Designer: Helen White
Consultant: Tom Collins,
Deputy Headmaster of St Leonards
CEP School, East Sussex

British Library Cataloguing in Publication Data

Woolfitt, Gabrielle
Air. – (The elements)
I. Title II. Series
551.51

ISBN 0 7502 0379 X

Typeset by White Design
Cover and inside artwork by Maureen Jackson
Printed by G. Canale & C.S.p.A. Turin
Bound in France by A.G.M.

CONTENTS

Words printed in **bold** are explained in the glossary.

WHAT IS AIR?

ABOVE **Nadu blows air into the special liquid to make bubbles.**

Air is all around us. Air is made of billions of tiny **particles** which move about all the time. These particles fill up spaces, such as the gaps between blades of grass or stones on a path.

The Earth's **atmosphere** is made of air. If the air were not there, then nothing could live. The Earth would be a lump of dead rock in space.

You can use your five senses to find out about the air.

How can you tell
that air is there?
You can see leaves being blown around
in an autumn breeze.
You can hear air whistling through small gaps
on a windy night.
You can smell flowers (and car exhaust gases!)
on a sunny afternoon.
You can taste salt
when you stand close to the sea.
You can feel air all around
when you fly your kite.
That's how you know
that air is there.

BELOW **The wind is blowing the foam on the sea. The children can feel the wind in their hair and taste the salty spray.**

GASES IN THE AIR

Air is a mixture of many gases. Scientists worked out what air is made of by doing experiments.
This is what they found out:

Name of gas	Amount in air	What the gas is used for
Nitrogen	78 %	Plants use it to make protein
Oxygen	21 %	Plants and animals need it to breathe and release energy from food
Carbon dioxide	0.03 %	Needed by plants to grow and make food

There are small amounts of other gases in the air too. The sign % means percentage. It tells you how many parts out of every hundred parts are that particular gas. If you could take exactly one hundred particles of air, you would find that seventy-eight of them were nitrogen particles.

Use a computer to plot some graphs to show how much of each gas is in the air. In the picture opposite, Toni and Ryan are plotting a bar chart showing the amount of gas up the side and the names of the gases along the bottom. A pie chart is also a good way of showing this information.

Can you think of any other ways to show which gases are in the air?

Gases in the Air

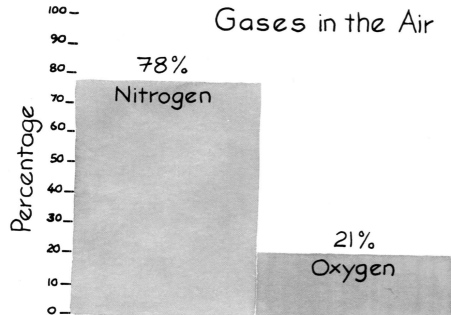

Percentage

100 —
90 —
80 —   78%
70 —   Nitrogen
60 —
50 —
40 —
30 —
20 —   21%
10 —   Oxygen   0.03%
0 —   Carbon Dioxide

Name of Gas

BREATHING

Use a stop-watch to count how many breaths you take in one minute. Can you work out how many breaths you take in a day?

ABOVE **Lindsay is counting her breaths. Jenny is timing one minute to help her.**

You don't have to remember to breathe. Your body does it **automatically**. You breathe in oxygen. You breathe out carbon dioxide.

All plants and animals need oxygen to make their cells work. Food and oxygen combine to release energy.

Cells give off carbon dioxide as a waste gas – just like a car giving off **exhaust gases**. The process of taking in and using oxygen, and giving out carbon dioxide, is called respiration.

Can you feel your ribs? Inside your ribs you have two lungs. When you breathe in, air is sucked into your lungs. Oxygen from the air goes into the cells of your body.

When you breathe out, your ribs squeeze your lungs. Unused air and waste carbon dioxide are pushed out.

BELOW **When you exercise you need more oxygen, so you breathe faster.**

SMELL

How can you tell that
...your dinner is being served?
...your aunt is wearing perfume?
...the people next door are having a bonfire?

BELOW **What do you think this Mexican food smells like?**

Smell tells us what is going on even when we cannot see what is happening. The air carries smelly particles from things right to our noses.

It is easy to know what a smell is. But it is very hard to describe in words.

Here are some words you can use to describe smells:

Sweet Fragrance

Heavy Perfume

Lemony Tang

HORRIBLE STINK

Nasty Pong

Strong Scent

List all the words you can find to describe smells. Write a poem about smells you like and those you hate.

RIGHT **Before you can speak or walk, you learn some very important smells. A baby can smell milk. It can also smell the people in its family and tell them apart from strangers.**

SPIRITS AND ANGELS

For thousands of years, people all over the world have believed in air creatures, such as fairies, angels, ghosts and spirits.

ABOVE **This is Tinkerbell, the fairy from the story of Peter Pan.**

Scientists have not found any real proof that spirits exist. That doesn't stop people from believing in them!

Nature spirits, like fairies, are supposed to help plants to grow and rivers to flow.

The Christian religion teaches that angels live in Heaven. They are God's messengers.

Some people believe that a ghost is the spirit of a person who has died in a horrible way. Henry VIII was an English king. He **executed** Anne Boleyn, one

LEFT **This Inuit mask shows a shaman's spirit flying free of its body. It is said that the spirit can fly to other parts of the world and find out what is happening there.**

of his wives. Some people say they have seen her ghost in Hampton Court Palace, where she died.

In some countries, evil spirits are blamed if a person becomes ill. A **shaman** or priest has to chase away the evil spirit before the person can be cured.

WINDS AND HURRICANES

BELOW **After Hurricane Gilbert there were huge floods. This plane was picked up by the force of the wind and dropped into the water.**

The wind is air moving from place to place. Have you ever licked your finger and held it up in the air on a windy day? One side of your finger feels colder than the other. The wind is coming from that side.

On a hot day, a gentle breeze helps to cool you down. Wind can also be very dangerous.

There are often very strong **hurricanes** in some parts of the world, such as the Caribbean islands. In 1988 Hurricane Gilbert destroyed buildings and hurt many people in Jamaica.

The windiest place on Earth is Commonwealth Bay in Antarctica. The wind blows faster than 320 km per hour several times each year!

Can you think of some ways of measuring how strongly the wind is blowing?

The hurricane which swept across southern England in 1987 ripped thousands of trees out of the ground. Buildings fell down and many people were left without electricity.

ABOVE **This car was crushed by a tree during the hurricane of 1987 in southern England. Hurricanes do not happen very often in Europe.**

ABOVE **These sailing-boats are racing in the sea.**

AIR POWER

People can catch the wind and use it to work machines. Sailing-boats have big sails which can be turned so they catch the wind. The wind pushes on the sail and makes the boat move.

Windmills used to be a common sight in Britain, Europe and the Middle East. Wind makes the sails turn round. The sails drive gears which make a huge stone turn. The stone grinds corn to make flour.

Modern windmills make electricity. Instead of turning a stone, the movement in the blades makes a **turbine** move round.

LEFT **This is a modern windfarm in California, USA. The wind turbines are made from metal.**

ABOVE **This Turkish windmill is like the first windmills, built over a thousand years ago.**

You can make a simple windmill out of stiff paper. Try to make a more complicated windmill that can be used to turn something.

Experiment with different kinds of sails. Can you use the wind to blow a skateboard along?

BIRDS AND BEES

The sun's rays
Lie along my wings
And stretch beyond their tips.
This Papago saying describes an eagle as it soars through the air.

Birds fly through the air. How many different kinds of birds can you think of? There are thousands!

Now make a list of flying insects. Remember that insects always have six legs and a body made of three parts.

Birds and insects can fly in three different ways. Sparrows and butterflies flap their wings to pull themselves through the air.

Eagles spread their wings and **glide** along on fast-moving streams of air called air currents.

Bees and humming-birds **rotate** their wings to push themselves along.

Watch some flying animals on slow film. How does each kind fly?

Look at a book about butterflies. With closed wings, a butterfly is hard to see. When it opens its wings you can see the beautiful colours and patterns.

The left side of a butterfly is the same as the right side. You can paint a picture of a butterfly.

Fold a large sheet of paper in half. Paint half a butterfly on one side of the paper. Then fold the paper over. When you open the paper, you will see a whole butterfly.

ABOVE **The male drone fly hovers in the air like a helicopter. He rotates his wings to stay in the same place. The female on the flower is watching him.**

RIGHT **The River Seine and the streets of Paris, seen from the Montgolfier hot air balloon.**

AIR TRAVEL

People have always wanted to be able to fly.

Leonardo da Vinci, an Italian artist and scientist, designed a type of helicopter in 1492. But nobody built one until 1923.

The first people ever to travel through the air were the Montgolfier brothers in France. They filled a big balloon with hot air from a burner.

Hot air is lighter than cold air so the balloon took off! The first passengers were animals.

The first aeroplane with an engine flew in 1903. Two Americans, Wilbur and Orville Wright, built it from wood and canvas.

Modern aeroplanes can fly at more than 1,600 km per hour. They are built from duraluminium, a very light kind of metal. Aeroplanes are pushed through the air by their engines.

Try making some paper aeroplanes in different ways and see which one flies the best.

Can you see how people learnt from birds when they made flying machines?

ABOVE **The Wright brothers testing out their aeroplane at Kitty Hawk in North Carolina, USA.**

AIR SPORTS

ABOVE **This is Tessa Sanderson, a British Olympic javelin thrower.**

When you throw a frisbee, it travels through the air using air currents. It moves easily because it is flat. The air passes above and below the frisbee.

A good javelin thrower knows how to use the movement of the air to make the javelin travel far. The record for a javelin throw at the 1988 Olympic Games was nearly 75 m!

Parachuting is exciting but dangerous. Before someone does a parachute jump, they need special training.

Imagine you are a **parachutist**. You go up in the air in an aeroplane. At a certain height you jump out. For the first few seconds you fall very fast. When you have moved away from the aeroplane, you pull the **rip-cord** and your parachute opens.

The parachute slows you down and you float down to earth. You still fall fast. People can break their legs if they land badly.

ABOVE **It takes years of training before someone can take part in parachute stacking like this. The parachutists have to be careful not to let their parachutes become tangled up.**

AIR POLLUTION

ABOVE **This lorry in Mexico City is giving off dirty fumes. Mexico City has very bad air pollution.**

If you walk along a main road during rush hour, you can smell exhaust gases from cars. These dirty gases cause air **pollution**. Sometimes air pollution is so bad that it kills living things.

The Earth may be getting warmer because of air pollution. When **fossil fuels** are burned to provide energy, carbon dioxide is given off. This gas traps some of the heat from the Sun inside the Earth's atmosphere, and so the Earth is slowly becoming warmer. This is called the 'greenhouse effect'. Find out why it could be so dangerous.

Burning a large amount of fossil fuels causes other problems. When coal is burnt in power stations, the waste gases go up into the air. Some of these gases **dissolve** in rain-water to make **acid rain**.

Acid rain damages many things. It can kill fish and other animals living in rivers and lakes. Acid rain harms trees too. It also wears away stone buildings.

BELOW **This forest has been destroyed by acid rain.**

KITES AND STREAMERS

RIGHT **These Japanese fish streamers are made from paper and then painted. They look very colourful against the winter sky.**

If you go to the park on a windy day you might see people flying kites.

Kites and streamers fly because the air pulls them up high. A good kite will fly in the gentlest of breezes. The best kites look great as well.

The shape of a kite is very important. A stretched diamond shape will fly well. The tail helps to balance the kite when it flies.

Kites should be light, but the material must be strong, too. Modern fabrics like rip-stop nylon are good because they do not tear. Fancy kites can be made from silk.

You can paint a paper kite. Use bright poster paints. A fabric kite can be dyed. You could write a message on it.

Fly your kite when it is not too windy. Keep away from power lines and trees.

RIGHT **These Malaysian boys have made this huge insect kite themselves. They are taking it to a kite festival.**

ACTIVITIES

Make a simple kite.

You will need:

two pieces of garden cane   string for tying, and flying the kite

paper, fabric or polythene   glue

1. Make the canes into a cross. Tie them together keeping the cross shape.
2. Cut a piece of paper or fabric 2 cm bigger than the cross all the way round. (A)
3. Glue the kite shape over the top of the canes.
4. Tie the flying string on to the central cross. This stops the string from slipping. (B)
5. Add a tail. Then decorate your kite. (C)

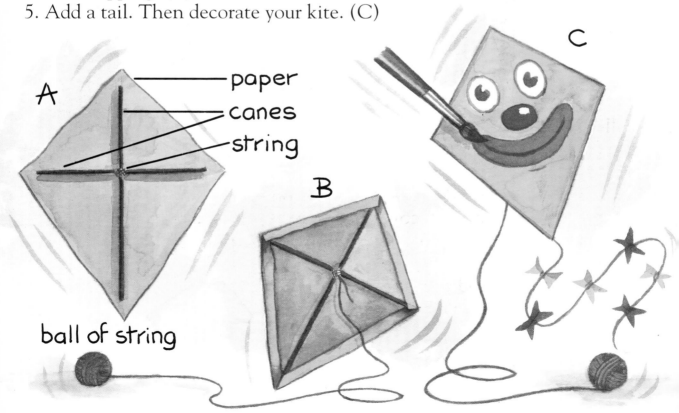

A

paper

canes

string

ball of string

B

C

Balloon messages

You can get to know people from another place by sending them a balloon message.

Write your name, address and message on the balloon. Or tie on a label, covered with plastic to stop it becoming wet. Climb to a high place to let your balloon go.

If you want to send your message further, ask an adult to fill some special balloons with helium. Helium is a very light gas. A balloon filled with helium can fly very high in the sky. It may travel for more than a thousand kilometres.

ABOVE **Lindsay and Jenny are deciding what to write on their balloons.**

GLOSSARY

Acid rain   Rain, snow and mist that has become acid from pollution.

Atmosphere   The mixture of gases that surrounds the Earth.

Automatically   Working or going by itself.

Dissolve   When a substance mixes with a liquid so you cannot see it any more.

Executed   Put to death.

Exhaust gases   The used gases given off by an engine.

Fossil fuels   Sources of fuel such as coal, oil and natural gas, which have been formed over thousands of years from the remains of dead plants and animals.

Glide   To be carried along by the air.

Hurricane   A violent storm with a very strong wind.

Papago   A North American native people.

Parachutist   Someone who jumps with a parachute from an aircraft.

Particle   A very tiny part of a substance.

Pollution   Anything that damages our environment. Pollution can spoil land, water and air.

Rip-cord   The cord that is pulled to make a parachute open.

Rotate   To turn round and round like a wheel.

Shaman   A medicine man.

Turbine   A machine with blades that are turned round by air, steam or water passing through them. A turbine makes electricity.

FINDING OUT MORE

Books
Animals, Animals ed. Laura Whipple (Hodder and Stoughton, 1990)
Atmosphere by John Baines (Wayland, 1991)
Breathing by Joan Gowenlock (Wayland, 1992)
My Science Book of Air by Neil Ardley (Dorling Kindersley, 1991)
Transport by Nigel Flynn (Wayland, 1991)

Music
The Sorcerer's Apprentice by Paul Dukas

Poetry
The Kingfisher Book of Children's Poetry ed. Michael Rosen (Kingfisher, 1985)
The Unicorn and the Lions ed. Moira Andrew (Macmillan, 1987)

Videos
The Sky by Edward Patterson Associates, 'Treetops', Cannongate Road, Hythe, Kent CT21 5PT, England
Why Doesn't Grass Grow on the Moon? by Viewtech Audio Visual Media, 161 Winchester Road, Brislington, Bristol, BS4 3NJ, England

Picture Acknowledgements
The publishers would like to thank the following for allowing their illustrations to be used in this book: Allsport 22; British Film Institute 12; Cephas (S. Hambrook) 27; Eye Ubiquitous (H. Lisher) 9; *cover* C. Fairclough; Hutchison (N. McKenna) 5,10, (J. Fuller) 17 (left); I. Lilly 4; Mary Evans Picture Library 20, 21 (right); Still Pictures (M. Edwards) 24; Swift Picture Library (R. Fletcher) 19; Tony Stone Worldwide (D. Smith) 23, 26; Topham 14; Tropix (V. Birley) 17 (right); Wayland Picture Library (A. Blackburn) 7, 8 and 29, (E. Miller) 15; Werner Forman Archive 13; T. Woodcock 21 (left); ZEFA 11, (W. Deuter) 16,18, (K. Goebel) 25.

INDEX

Page numbers in **bold** indicate subjects shown in pictures, but not mentioned in the text on those pages.